Water

217

沙漠中的水坝

Dams in the Desert

Gunter Pauli

[比]冈特·鲍利 著

[哥伦]凯瑟琳娜·巴赫 绘

闫世东 译

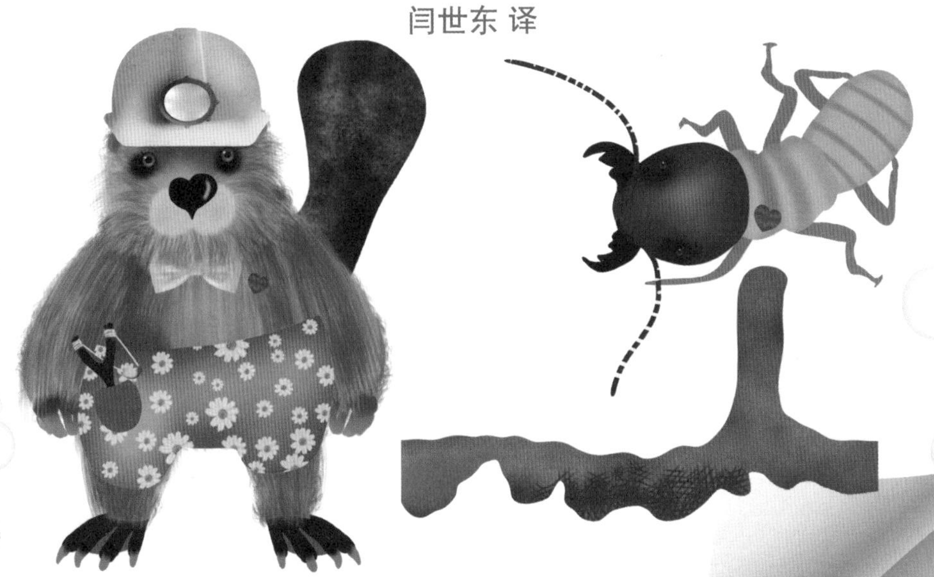

上海遠東出版社

丛书编委会

特别感谢以下热心人士对童书工作的支持：

匡志强　方　芳　宋小华　解　东　厉　云　李　婧
刘　丹　熊彩虹　罗淑怡　旷　婉　杨　荣　刘学振
何圣霖　王必斗　潘林平　熊志强　廖清州　谭燕宁
王　征　白　纯　张林霞　寿颖慧　罗　佳　傅　俊
胡海朋　白永喆　韦小宏　李　杰　欧　亮

目录

Contents

一只河狸第一次游览纳米比亚。他知道这里有世界上最古老的沙漠。当他遇见一只沙漠白蚁，他问道：

“你能向我解释一下这些随处可见的神秘的圆圈是什么吗？”

“你是说这些围绕着一片片贫瘠沙地的草环？”

A beaver is visiting Namibia for the first time. He knows that it is home of the oldest desert in the world. When he meets a sand termite, he asks:

"Could you please explain to me what these mysterious circles are that I see everywhere?"

"You mean these rings of grass, around the barren patches of sand?"

……河狸游览纳米比亚……

... beaver is visiting Namibia ...

……龙的足迹！

... dragon's footprints!

“是的。我听说过一些关于草环起源的奇怪设想，有些人甚至把它们想象成龙的足迹！”

“如果是龙造成了这一切，那么我就是那条龙！我们白蚁以这种方式为所有共同生活在这片旱地的生物带来生机，从壁虎、鼹鼠、蜘蛛到猎捕它们的胡狼。”

“是的，正是你那出色工程师的名声使我慕名而来，看看你如何建造有‘空调’的蚁穴，那些能控制温度和湿度的蚁穴。但我没有看到你们那出名的烟囱。”

“Yes. I’ve been hearing some strange ideas about their origin, with some people even imagining they are a dragon’s footprints!”

“If these were made by a dragon, then I am that dragon! This is our termites’ way of bringing life to all who share this dry land with us, from geckos, moles, and spiders, to the jackals that hunt them.”

“Yes, and it is your fame as a great engineer that brought me here, to come and see how you build your air-conditioned mound, where you always have the moisture and temperature under control. I don’t see any of your famous chimneys though.”

“我们建造了一个小型地下绿洲，而不是在地上建造烟囱。”

“我从未听过地下绿洲。但请你告诉我，白蚁，如果你不建造烟囱，你还是真正的白蚁吗？”

“为什么每一个白蚁巢穴都要有一个烟囱呢？我们挖小隧道，咀嚼木材，使空气循环起来，还收集如果在地表就会蒸发掉的地下水，这样就可以起到相同的作用。生命也因此而繁荣。”

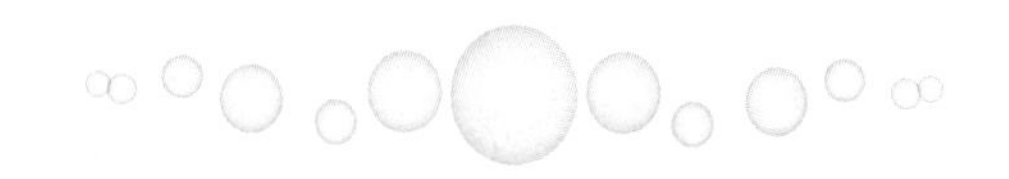

“Instead of creating a chimney above the ground, we create a tiny oasis underground.”

“I have never heard of an underground oasis. But tell me, Termite, if you don’t build chimneys, are you a real termite?”

“Why does every termite nest have to have a chimney? We get the same result, when we dig small tunnels, chew up wood, circulate the air, and trap the water underground that would’ve evaporated above. Thus life thrives.”

……我们建造了一个小型地下绿洲。

…we create a tiny oasis underground.

……几千个圆圈……

…how many thousands of them there are…

“确实，这是了不起的工程。但每个圆圈只覆盖了如此小的一片区域。你这样是没有效果的。”

“哦，请看看这几千个圆圈吧。每个圆圈都是由数百万勤奋的‘工人’建造的，有着生机勃勃的地下环境。”

“生机勃勃？我所看到的，是你把小圆圈底下的土壤变得肥沃，这使得一些植物可以在通常难以生长的地方成长。植物吸引了一些食草动物，而动物们则留下了排泄物。太聪明了！但我仍不认为这算是生机勃勃。”

Great engineering, true. But each circle covers such a tiny area. You have no impact!”

“Well, just look at how many thousands of them there are. And each one has such a lush environment underneath, created by millions of very hard workers.”

“Lush? All I see, is that you turn a tiny circle under the ground fertile, so a few plants can grow where normally nothing would grow. This attracts a few grazers, and they then leave their droppings behind. Clever! But I still wouldn’t call it lush…”

“这取决于你拿它同什么比较。这里确实是生机勃勃的环境——如果你唯一的比较对象是炙热的沙地的话！现在你不正在通过建坝把河流变成湖泊吗？”

“是的，我们的确用当地的树建造木制结构体来挡住水流。水流缓缓地渗出，留下我们从来不吃的大型鱼类。我们吃树皮、嫩枝和树叶，给河边低洼地施肥。”

“我很好奇，你有没有设想过在沙漠中建造水坝呢？”白蚁问道。

“It all depends on what you compare it with. It is very lush – if your only other option is the hot sand! Now aren't you the one who turns rivers into lakes by building dams?”

“Yes, we do build wooden structures from local trees, to retain the water. The water slowly seeps through, leaving behind the big fish, which we never eat. We eat bark, twigs and leaves, fertilising the meadows.”

“I wonder, have you ever thought about building a dam in the desert?” Termite asks.

……用当地的树建造木制结构体……

… wooden structures from local trees …

……蚁狮……

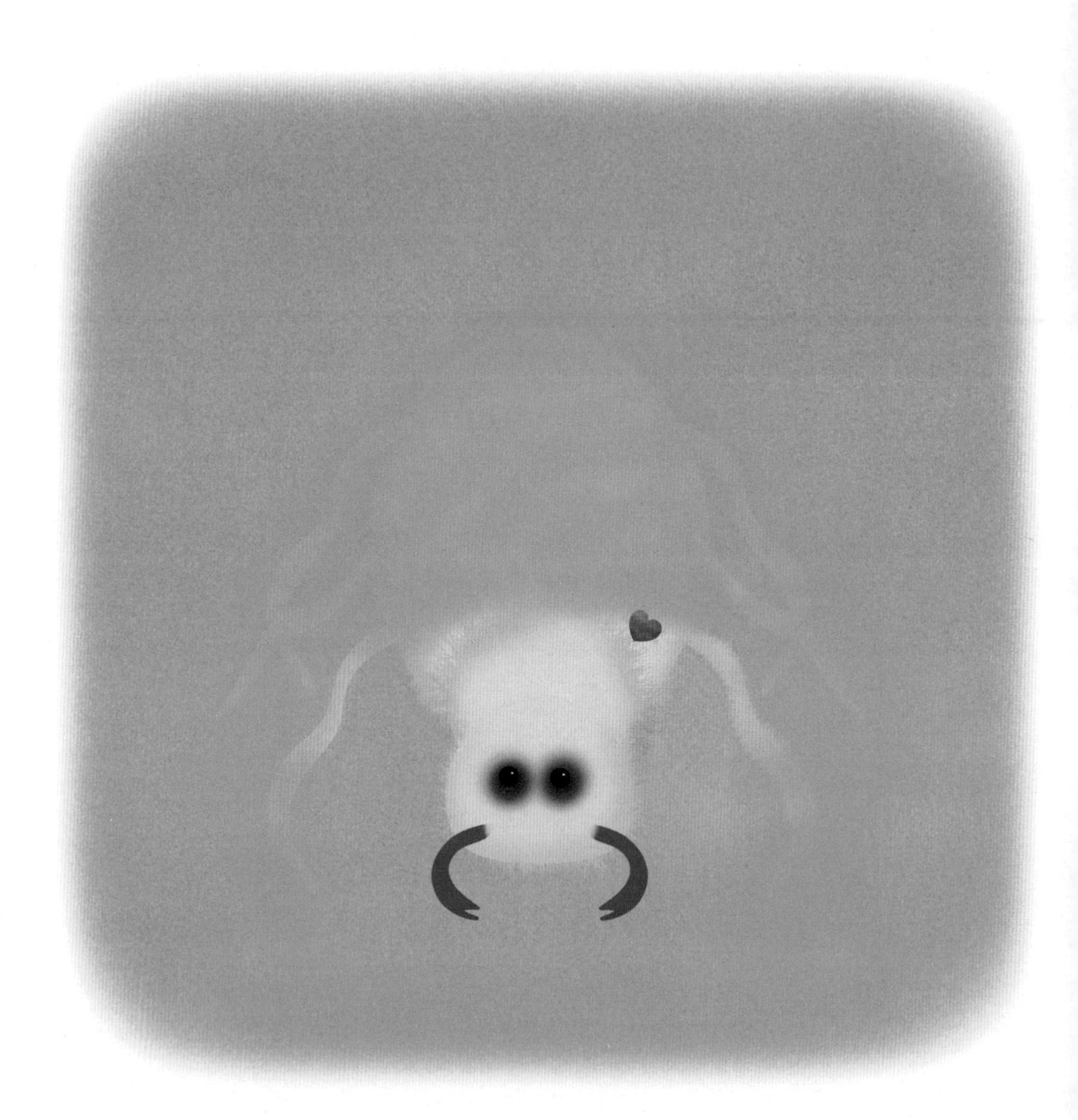

… the ant lion …

“在沙漠里？这怎么可能？建坝首先需要的是水。”

“好吧，也许你能够想象出遵循同样原理并流动的东西？我们有一种昆虫，蚁狮，为了捕获大型猎物可以让沙子流动起来。”

“我未曾听说过这件事，但我明白了，细小的沙粒可以像水一样流动。数世纪以来，沙钟已经告诉信徒们何时祈祷。唯一的问题在于沙子中没有鱼！”

“In the desert? How could you do that, when the first thing you need is water?”

“Well, perhaps you could imagine something fluid that works along the same principles? We have an insect, the ant lion, that makes sand flow like a fluid to catch big prey.”

“I’ve never heard of it, but I do get the logic, of fine grains of sand flowing like water. For centuries, sand clocks have told believers when it is time to pray. The only problem here is that there are no fish in the sand!”

“其实，沙子中是有鱼的。但我更希望种一些草以吸引食草动物，然后利用它们的排泄物给土壤施肥！”

“好主意！与我们所做的差不多，只是没有水。现在，如果你想捕鱼，你需要一条河。我在方圆几英里内看不到水源。你有什么存贮水的方法吗？”

“当然！”

“Well, we do have sand fish. But I’d rather grow some grass for the grazers, and have them fertilise the soil with their droppings!”

“Good point! And comparable to what we do, only without any water. Now, if you want to trap fish, you need a river. And I don’t see any water for miles around. Any idea how to get hold of any?”

“Oh yes!”

……种一些草以吸引食草动物……

... rather grow some grass for the grazers.

……最出名的水利工程师……

… most celebrated water engineers …

“你打算表演一段神奇的祈雨舞吗，还是通过更科学的方法？”

“当然不是跳舞。”白蚁回答道。“别忘了，我们是水利工程师。”

“你们可以自称是全世界的沙漠中最好的水利工程师，但我们是森林中最出名的水利工程师。”

“Are you going to perform a magic rain dance, or is it something more scientific?”

“Absolutely no dancing,” Termite replies. “Do not forget, we are the water engineers.”

“You may pretend to be the best water engineers in the world right out here in the desert, but we are the most celebrated water engineers of our forests.”

“大概在有河的地方，你们是杰出的工程师，但在没有河的地方，我们才是杰出的工程师。”

“我很好奇，你是在开玩笑还是说你非常不切实际？你真的有一些独特的东西可以分享吗？”河狸问道。

“我当然有！当人们还在为产生沙漠中圆圈的原因而争论不休时，我们正在为了大家的利益不断工作，并为此充分利用现有的东西。”

“我必须承认，白蚁，你不仅是一个大胆的工程师，还是一个伟大的梦想家。”

……这仅仅是开始！……

“And you may be great where there are rivers, but we are great where there are no rivers.”

“Are you pulling my leg, or are you just a dreamer, I wonder. Do you really have something unique to share?” Beaver asks.

“Of course I do! While people are endlessly debating what causes the circles in the desert, we are working away at making the best use of what we have, for the benefit of all.”

“I must admit, Termite, you are as great a dreamer as you are a daring engineer.”

... AND IT HAS ONLY JUST BEGUN!...

……这仅仅是开始！……

... AND IT HAS ONLY JUST BEGUN! ...

Did You Know ?

你知道吗？

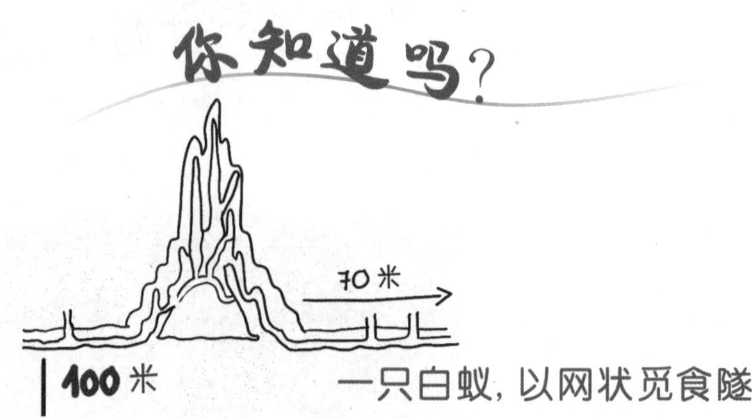

A termite's range of influence can be as far as 70 metres from the nest, in the form of a network of foraging tunnels, and as deep as 100 metres, in the form of vertical transport of water and soil. The circles can measure up to 12 metres across.

一只白蚁，以网状觅食隧道的形式对巢穴外最远 70 米处产生影响，并以水分和土壤的垂直输送系统的形式对最深 100 米处产生影响。圆圈最大直径可达 12 米。

Termites are harvested and prepared as food. In Nigeria, the Igbo people refer to it as aku mkpu, Akwa-ibom people refer to it as ebu, Yorubas call it esusun, while the Hausa people call it khiyea. Flying termites are collected, and steamed or roasted to be eaten.

人们采集并烹制白蚁作为食物。在尼日利亚，伊博族人把白蚁叫作 aku mkpu，阿夸依博姆地区的人把白蚁叫作 ebu，约鲁巴族人把白蚁叫作 esusun，而豪萨族人把白蚁叫作 khiyea。人们收集会飞的白蚁，蒸煮和烘烤后食用。

Termites, with strong and proven antibacterial properties, are used in traditional remedies (especially in south India), for treating diseases associated with microorganisms. Termitaria and surrounding soil have greatly reduced presence of many bacterial strains.

白蚁有着强力且经过证明的抗菌特性，在传统疗法（特别是在印度南部）中被用于治疗与微生物有关的疾病。在白蚁巢和周围的土壤中，许多细菌的数量大大减少。

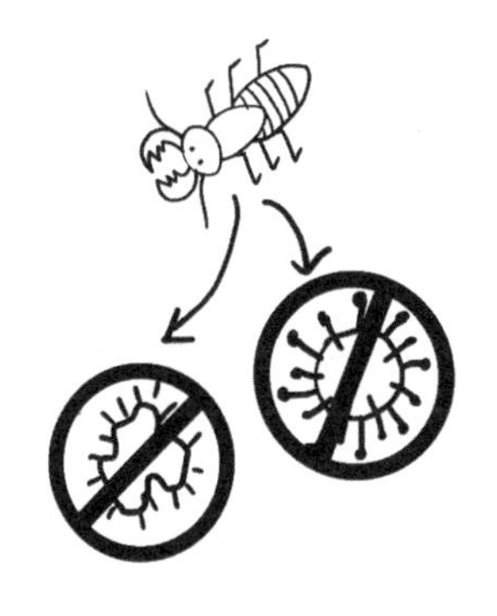

Termites contain antiviral properties, which have been proven effective for treating colds, sore throats, coughing, hoarseness, asthma, catarrh, bronchitis, influenza, whooping cough, sinusitis and tonsillitis.

白蚁具有抗病毒特性，已被证明可有效治疗感冒、喉咙痛、咳嗽、声音嘶哑、哮喘、卡他性炎、支气管炎、流行性感冒、百日咳、鼻窦炎和扁桃体炎。

Peptides from termites are effective against selected fungal attacks. Termites, grow their own mushrooms on partly digested detritus, promoting the growth of one unique species of mushrooms, while stopping the growth of others.

白蚁分泌的肽对抵御特定真菌的攻击有效。白蚁在部分消化的碎屑上种植自己的菌菇，促进一种独特菌菇的生长，同时阻止其他菌菇的生长。

Termites offer multiple ecosystem services. Their termitaria contain a high amount of phosphorous and nitrogen that support plant growth by recycling waste matter such as decayed wood, plant material and cellulose, as well as faeces and dead animals.

白蚁在生态系统中具有多种功能。白蚁巢中含有大量促进植物生长的磷和氮。白蚁通过回收废弃物如腐烂的木材、植物材料、纤维素、粪便和动物尸体来获取这些磷和氮。

About 3,000 termite species have been classified, of which one-third is found in Africa, with its extremely abundant termitaria in certain regions, such as in the northern Kruger National Park, (South Africa), with about 1.1 million active termite mounds.

人们已经对大约 3 000 种白蚁物种进行了分类，其中三分之一在非洲发现，而且在某些地区白蚁巢穴极为丰富，例如在南非克鲁格国家公园北部，大约有 110 万活跃的白蚁丘。

Termites are hard-working insects that form unique colonies that contribute to the rehabilitation of ecosystems, through aeration of the soil, replenishing the area with nutrients and controlling viruses, bacteria and fungi. They are an inspiration to many.

白蚁是勤劳的昆虫，它们形成独特的聚落，通过为土壤通气为该区域补充营养，控制病毒、细菌和真菌生长，为生态系统的恢复作出贡献。白蚁鼓舞了许多人。

Think about It

Do termites sound appetising to you?

白蚁会让你产生食欲吗?

Can you imagine how termites create an underground water supply in the desert?

你能想象白蚁如何在沙漠中创造地下水源吗?

Is the beaver good for the ecosystem where it is cutting down trees?

河狸砍伐树木,对生态系统有好处吗?

Is water only required in large volumes like rivers and dams, or is it also very helpful as moisture in the soil?

是否只有大量的水才是有用的,比如河流和大坝这样的地方的水?还是说微量的水也是有用的,比如土壤里的水?

Do It Yourself!

自己动手！

Would anyone even consider a termite as a favourite animal, or a pet? Not likely, is it? So, let's investigate the reasons for people having an aversion to these incredible insects. Question people about their misgivings and compile a list of their reasons for these. Now draw up a list of everything that makes these insects so unique and so important to ecosystems. Armed with these facts, start a discussion with your friends and family members, one that will convince them that the termite is a great partner in building sustainability.

有人会把白蚁当作喜爱的动物或者是宠物吗？不太可能，对吗？因此，让我们调查一下人们厌恶这些令人难以置信的昆虫的原因。询问人们的担忧之处并汇编成原因清单。现在，列出所有使这些昆虫如此独特且对生态系统如此重要的一切原因。有了这些事实支撑，就可以与你的朋友和家人讨论，使他们相信白蚁是促进可持续发展的可靠伙伴。

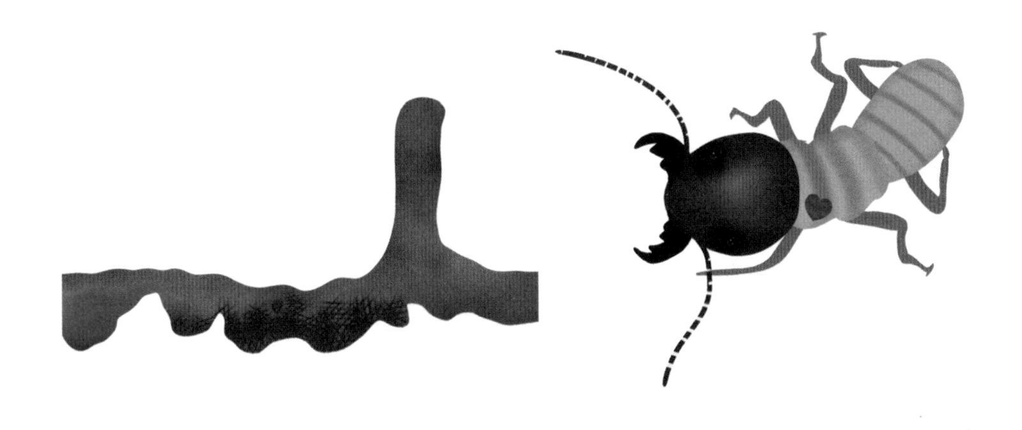

学科知识

Academic Knowledge

生物学	白蚁与蟑螂都被归为蜚蠊目；白蚁在地下建立了茂密的真菌花园；多齿白蚁在地下不超过一米的地方建立了多个没有烟囱的蚁巢，这些蚁巢通过地下隧道相连；河狸是植食性啮齿动物；河狸的社会组织结构：一夫一妻，还有幼崽；白蚁利用共生的披发虫消化木材，披发虫世代生活在白蚁的肠道中。
化　学	高岭土是在白蚁巢穴中发现的药用黏土，对治疗胃部疾病非常有用；蚁丘含有大量的钙和铁；在稀树草原中，色素沉着可以保护白蚁免于脱水并抵抗紫外线；河狸香和蓖麻油。
物　理	白蚁化石仅在琥珀和页岩中出现，因为它们的身体柔软且会迅速腐烂，然而蚁丘和隧道可以变成化石；超细沙粒像液体一样，会变形；沙钟中的沙流指出精确的时间。
工程学	蚁巢的“甲壳”由硬质的胶结砂外层组成，蚁巢中央的生活空间表面光滑，呈拱形且平坦；由有机和无机材料的混合物制成的球形地下蚁巢，坚实程度相当于可折叠硬纸板。
经济学	欧洲河狸因自身产生的河狸香而被猎杀至灭绝，河狸香被用于止痛、消炎和解热；人们用河狸香生产香水，并用作调味品的增强剂。
伦理学	加强共同利益；为了使白蚁能够生存，整个生态系统都需要蓬勃发展；从他人的角度看待现实的能力。
历　史	白蚁是在侏罗纪或三叠纪从蟑螂进化而来的；化石记录表明，2 000万年前非洲就有白蚁；北半球的河狸出现于始新世；1960年从加拿大引入到火地岛（阿根廷和智利）的河狸被认为是一种入侵物种。
地　理	旱地生态系统占全球陆地面积的40%，为世界六分之一的人口提供支持；纳米布沙漠是早中新世末期形成的，“年龄”大约为1 700万年，是最古老的沙漠。
数　学	白蚁蚁后每天产下4 000个卵，可存活100年；同一个蚁巢可以废弃，也可以继续使用1 000年（在天气允许的情况下），因此白蚁灭绝的概率为零；蚁狮的圆锥形陷阱的几何形状。
生活方式	看法取决于人的视角和背景，因此我们应考虑那些处于不同参照系的人的观点；我们试图简化因果关系。
社会学	纳米比亚的丛林人认为“精灵怪圈”具有魔法和精神力量；一些游牧民族，在没有其他牧草的时候，会用“精灵怪圈”附近的草喂养牲畜。
心理学	白蚁印证了一句格言：一个人走得快，一群人走得远；团队合作在成员彼此信任，具有合作的经验并同情彼此时才有效。
系统论	通过降雨情况、生物数量、季节温度以及白蚁的存在可以精准地预测出“精灵怪圈”的存在；旱地生态系统的人口压力和土地利用集约化导致许多类似地区严重的土壤退化和荒漠化。

情感智慧

Emotional Intelligence

河 狸

河狸千里迢迢赶来向控制温度和湿度的出色工程师——白蚁学习。他从未听说过地下绿洲，这令他惊讶。当他看到白蚁的工作并掌握其原理时，他公开表示他认为“精灵怪圈”太小，无法对环境产生影响。当受到质疑时，他已经有所准备并乐于分享。当白蚁让他想象在沙漠里做他在家里做过的事情时，他心态失衡了。他在寻找线索，但由于没有水也没有河流而感到绝望。他自信地宣称自己是最著名的水利工程师，但事实上这种说法仅适用于森林。他越来越困惑，但最后他对白蚁敢于深度思考的能力深表钦佩。

白 蚁

白蚁说她制造圆圈的主要动机是为大草原带来生机。她分享了自己的适应方法，并说明她不使用烟囱的原因。当河狸奚落“精灵怪圈”直径太小以至于影响很小时，她要求河狸考虑圆圈的数量。随后她提出在沙漠中修建水坝的想法。白蚁坚持认为必须要有创造更多水的方案，而且这些方案应突破传统的因果关系研究。她充满信心地指出使生态系统再生需要一些大胆的想法。

艺术

The Arts

让我们用艺术方法从两个不同的角度来诠释“繁茂”一词。沙漠中一处植被生长繁茂的地点，也就是白蚁负责的这些圆圈，与周围的沙地相比形成了一种茂密的景观。接下来，绘制一片茂密的热带雨林，这符合我们对“繁茂”一词的传统定义。要注意选择合适的颜色来展示两种不同的“繁茂”。现在，你已经从两个不同的角度掌握了“繁茂”的定义。

思维拓展

Systems: Making the Connections

白蚁的社会组织清楚地表明了它们维持和恢复所有生命共同依赖的生态系统的能力。白蚁的系统性方法，即利用物理定律控制湿度和温度以及对如何触发和影响养分循环的清晰理解，使白蚁成为维持和恢复土壤肥力的关键物种。世界人口激增，特别在像稀树草原这样的地区，那里居住着全世界40％的人口，沙漠化也在不断加剧。随着土地生产压力越来越大，我们必须学会表现得更好。种植更多的树木不足以恢复土地的肥力。我们不合理地期望地球产出超出其能力范围的资源，并继续开发稀缺的水和土壤资源。结果，现有的生产系统受到损害，并且越来越倾向于崩溃。我们正在挖取资源，最糟糕的是，我们还在与几种物种进行斗争，这些物种对于维持整个系统可能的最大生产力至关重要。逆转目前的状况首先需要一些大胆的想法和建议。接下来，我们要从不同于那些在四季分明、有丰沛降雨和茂密森林的环境中长大的科学家和创想家的视角去看待现实。他们的参照系是温带气候，而复原快要变成荒漠的大草原则需要采取截然不同的方法。在这个寓言中，白蚁促使它的访客河狸去看河，或者想象一下如何在没有水的广阔地区创造河。致力于实现共同利益并立刻作出改变的人们提出的创意经常被视为“做白日梦”或“开玩笑”。一个想法听起来可能很疯狂，但可能会促进工程学上的大胆创新，并找到以前无法想象的方案。白蚁在干燥而复杂的环境中生存了超过2 000万年。当然，了解它们如何生存并成功度过冰河时代，可以使我们更全面地探寻满足所有人需求的未来。

动手能力

Capacity to Implement

你要如何将一块贫瘠的土地变成一块充满生机的土地？想象一下，将贫瘠的土地变成具有生产力的生态系统。你可以向白蚁学习，这非常有用。列出你认为转变所需的前提条件。现在我们来看一些成功地将沙漠和稀树草原系统转变为具有生物多样性的生态系统的案例，比如哥伦比亚的拉斯加维奥塔斯，以及中国的黄土高原。现在加上你从白蚁身上学到的知识，制定出实现转变的计划，并与他人分享你的看法。

故事灵感来自

This Fable Is Inspired by

尼可·巴格

Nichole Barger

尼可·巴格最初在美国印第安纳州布卢明顿的印第安纳大学学习，然后于 1995 年在华盛顿州奥林匹亚的长青州立大学获得环境科学学士学位。1998 年，她在加利福尼亚大学伯克利分校获得了生态学硕士学位。最终，她完成了一项研究计划，并于 2003 年在科罗拉多州柯林斯堡的科罗拉多州立大学获得了生态学博士学位。她目前在科罗拉多州博尔德的科罗拉多大学生态与进化生物学系工作。她是干旱土地生态学家，致力于更好地了解气候变化和土地利用对植物群落的影响。她的研究连接了群落生态学、生态系统生态学和景观生态学。她的研究涵盖了陆地植物生态学、土壤生物地球化学和树木年代学等领域，以进一步了解干旱地区生态系统的结构和功能以及生态系统的可持续恢复和管理。

图书在版编目(CIP)数据

风特生态童书.第七辑:全36册:汉英对照/
(比)冈特・鲍利著;(哥伦)凯瑟琳娜・巴赫绘;
何家振等译.—上海:上海远东出版社,2020
ISBN 978-7-5476-1671-0

Ⅰ.①冈… Ⅱ.①冈… ②凯… ③何… Ⅲ.①生态
环境-环境保护-儿童读物—汉英 Ⅳ.①X171.1-49

中国版本图书馆CIP数据核字(2020)第236911号

策　　划　张　蓉
责任编辑　祁东城
封面设计　魏　来李　廉

冈特生态童书
沙漠中的水坝
[比]冈特・鲍利　著
[哥伦]凯瑟琳娜・巴赫　绘
闫世东　译